Let's Investigate
Soft, Shimmering Sand

Madelyn Wood Carlisle

Illustrations by Yvette Santiago Banek

BARRON'S

All inquiries should be addressed to:
Barron's Educational Series, Inc.
250 Wireless Boulevard
Hauppauge, NY 11788

International Standard Book No. 0-8120-4972-1

Library of Congress Catalog Card No. 92-44756

Library of Congress Cataloging-in-Publication Data

Carlisle, Madelyn.
    Let's investigate soft, shimmering sand / Madelyn Wood Carlisle ; illustrations by Yvette Santiago Banek.
        p.   cm.
    Includes index.
    ISBN 0-8120-4972-1
    1. Sand—Juvenile literature. [1. Discusses what sand is, how it is formed, how beaches and dunes are built up, how concrete and glass are made from sand, and how animals and plants survive in the desert. 2. Sand.] I. Banek, Yvette Santiago, ill. II. Title.
QE471.2.C38   1993
553.6'22—dc20   92-44756
                                                                                            CIP
                                                                                            AC

PRINTED IN HONG KONG
3456   8800   987654321

Contents

Sand Everywhere

There is a lot of sand on our planet. Geologists have figured out that if all the sand on Earth could somehow be dumped on the United States, it would cover the entire country with a layer 3 miles (5 kilometers) thick!

Much of the world's sand is found on beaches and on the continental shelves, those underwater areas close to shore. There are also vast deposits of sand far out in the oceans, on the deepest ocean floors. Many of our rivers and streams have sandy bottoms. The desert regions of Earth are not all sandy, but where they are there are sometimes giant sand dunes hundreds of feet high.

There are also huge deposits of sand buried deep in the ground. Some of these underground deposits are mined for minerals or drilled into for oil.

When you think of sand, you probably think of loose sand, the kind you played with in your sandbox when you were younger, the kind you like to walk on at the beach. But not all sand is loose sand, the kind you can pick up and let sift through your fingers.

Sand that has been compressed under water for a long time gets packed together very tightly. Then, if the packed sand becomes exposed and dries out, it turns into a hard rock called sandstone.

When sandstone is eroded by wind, rain, and other sand particles blowing against it, it is sometimes formed into beautiful and fascinating shapes. The pillars, arches, and natural bridges you can see in some places in the American Southwest are made of eroded sandstone that was once buried beneath the sea.

It isn't just nature that builds with sandstone. Some of the oldest structures built by humans were made with blocks of sandstone or were carved out of sandstone cliffs.

Your life would not be the same without the many things we have learned to make with sand. Later in this book you will read about some of them.

Much of the world's sand is found on beaches, like those on Lake Superior (top), and in North Carolina (center). Sand can also turn back into rock. In the American West, there are many beautiful formations made of sand turned to stone (bottom).

The sands of Saudi Arabia and other Middle Eastern countries stretch as far as the eye can see.

What Is Sand?

Have you ever picked up a handful of beach sand and looked at it through a magnifying glass? On the beach it may have looked light brown, but when you study the magnified grains in your hand you see that they are really many different colors. Some are brown, all right, but others may be white, pink, orange, red, coral, green, or black.

This may make you wonder, "Just what is sand?"

Geologists use the word sand to describe earth particles of a certain size. Grains of sand are larger than $\frac{1}{400}$ of an inch (0.06 millimeters) and smaller than $\frac{1}{12}$ of an inch (2.1 millimeters). Particles smaller than sand grains are called silt; larger particles are called gravel.

But what are sand particles made of?

The most abundant element on Earth is oxygen. The second most abundant element is silicon. Silicon is present everywhere in the Earth's crust. Oxygen and silicon combine to form silica, the main ingredient of sand.

Take some sand between your fingertips and rub the grains together. You will find that what seemed soft when you walked on it is actually made up of very hard particles. Each small grain is really a tiny stone. Most are pieces of quartz. Quartz is a very common mineral made of silica. When it is pure it is colorless and transparent. That is one of the reasons why sand seems to shimmer in the sunlight.

Once, maybe millions of years ago, the little grains of sand you are holding were parts of much larger rocks, maybe gigantic boulders. The big rocks may have fallen off cliffs and shattered. Pieces may have landed in streams where the water tumbled them about. They bumped against other rocks and their rough corners and edges were chipped off. They may have traveled many miles, and for many years, in a river on its way to the sea.

By the time those pieces of rock reached the ocean, they had probably become small pebbles or coarse gravel. And there, on the beach, the waves and the winds tossed them back and forth, back and forth, wearing them down still more.

The original large rock didn't have to fall into a stream to be turned into sand. It might have been worn down by rain, snow, and ice. Eventually it became small enough so that it could be picked up by the wind and blown about. Hitting against each other or scraping along the ground for thousands, maybe millions, of years, the rocks became smaller and smaller until, finally, they were just tiny grains of what we call sand.

In the handful of sand you looked at through the magnifying glass, you may have seen that the grains were not all the same size or shape. Some of the smallest grains were perfectly round, whereas larger ones were irregular and had sharp corners. The smaller, rounder grains had been traveling and tumbling about longer than the more roughly shaped ones. They had become what geologists call mature sand.

In this sandstone cliff in Zion National Park, in Utah, you can clearly see the layers of sand that were deposited at different times in Earth's history.

Some of the most colorful sand in the United States is in Utah's Coral Pink Sand Dunes State Park.

Sandy Beaches

Whether it is at a bend in a river, the shore of a lake, or the edge of the ocean, a beach is a wonderful place. Have you ever known anyone who didn't like to go to the beach?

You can have fun on any beach, even one that is pebbly and hurts your feet, but you can have the most fun of all on a beach that is sandy. There are many beautiful sandy beaches in the world. Maybe the one you know best has sand that is tan, gray, or almost white, but beaches come in many colors.

In Hawaii there are some black sand beaches. If you guessed that the black sand particles are really tiny bits of volcanic rock, you guessed right.

Bermuda is famous for its pink beaches. The color comes from the bits of seashell and coral that are mixed in with the white sand.

Next time you spend a day on a sandy beach, you might like to think about why that wide strip of sand is there. Where did the sand come from?

The beaches on the island of Bermuda get their rosy color from tiny bits of rose-colored seashells mixed in with the white sand.

As you have already read, the sand particles may have traveled for hundreds, maybe thousands, of miles, and for many thousands of years, to get to where they, along with millions of other sand grains, gradually made a beach.

Of course, the sand grains you see on your beach now are not the same ones that first stopped there to make a beach. They may not even be the same grains you walked on last time you were there. This is especially true of ocean beaches, where high tides and the big waves of stormy seas can sometimes change the shape of a beach in a few hours.

Even calm seas, with only small ripples lapping at the shore, constantly change the sand on a beach. Watch the water as it meets the shore. Each time a wave washes up onto the sand, it carries new grains of sand with it. Some are left there on the shore; others are carried back out into the sea. Small, gentle waves tend to leave more sand than they take away. Big, powerful waves do just the opposite, removing more sand than they bring in. That is why, after a fierce storm, beaches can sometimes be much smaller than they were before.

You have heard about endangered animals and endangered plants. Did you know that Earth's beaches are endangered too? In many places in the world, beaches are getting smaller and smaller. Some have even entirely disappeared. Scientists say that along the Atlantic Coast of the United States the beaches are losing an average of 3 feet (1 meter) a year. The beaches on the Gulf Coast of Texas are losing even more.

Yet winds and waves are not the greatest destroyer of beaches. People do far more damage to beaches than do the most violent storms. People who have homes close to the shore often think they can protect the beach by building walls in front of their properties. The walls interfere with nature's way of keeping the beach supplied with sand. One little strip of beach may be protected for a while, but other beaches down the shore can find their supply of new sand completely cut off. Soon, instead of sandy beaches, there are only stony ones.

NATIONAL SEASHORES

Many of the biggest and sandiest ocean beaches in the United States have been set aside as National Seashores. Visit these places and you can see some of America's most beautiful beaches:

- Cape Cod National Seashore, in Massachusetts. 43,526 acres (17,615 hectares).
- Fire Island National Seashore, in New York, off the south shore of Long Island. 19,579 acres (7,924 hectares).
- Assateague Island National Seashore, off the coasts of Maryland and Virginia. 39,631 acres (16,039 hectares).
- Cape Hatteras National Seashore, on barrier islands off North Carolina. 30,319 acres (12,270 hectares).
- Cape Lookout National Seashore, on barrier islands off North Carolina. 28,415 acres (11,500 hectares).
- Cumberland Island National Seashore, on an island off Georgia. 36,410 acres (14,735 hectares).
- Canaveral National Seashore, off Florida's Atlantic Coast. 57,627 acres (23,322 hectares).
- Gulf Islands National Seashore, in the Gulf of Mexico, off Florida and Mississippi. 65,817 acres (26,636 hectares).
- Padre Island National Seashore, on an island off the Gulf Coast of Texas. 130,697 acres (52,893 hectares).
- Point Reyes National Seashore, on a peninsula north of San Francisco, California. 71,046 acres (28,752 hectares).

Sand Dunes

At the back of your favorite beach, there may be dunes—low hills of sand shaped into beautiful curves by the wind. Just like the flat beach nearer the water, dunes are also fun to play on. Next time you look at a sand dune, stop to think about how dunes are made, why they are there, and what keeps them from just blowing away.

For a sand dune to form, three things are needed: a source of sand, wind that picks up the sand and carries it away, and an obstacle that keeps the sand from going any farther.

The highest sand dune in North America is in Great Sand Dunes National Monument, in Colorado. It is 700 feet (213 meters) high. The dunes there spread over 55 square miles (142 square kilometers). Where did all that sand come from?

Sixty miles (97 kilometers) to the west are the San Juan Mountains. The winds, which blow from west to east, are constantly carrying with them tiny particles worn away from the San Juans. Sometimes the particles are airborne; more often they are blown along the ground, maybe just a fraction of an inch at a time. Either way, they finally come to another mountain range, this time the mighty Sangre de Cristos.

The arrows indicate the direction of the wind and show how it shapes sand into different kinds of dunes. Top left: transverse dunes; top right: barchan dunes; bottom left: seif dunes; bottom right: star dunes.

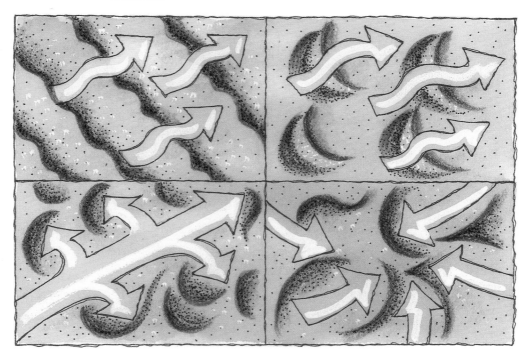

The different winds that swept over this vast area of dunes sculpted the sand into many strange and beautiful shapes.

In some parts of the world, farms, fields, houses, even entire towns, have been buried by creeping dunes.

The mountains don't stop the wind. It sweeps up the western slope and on over the top of the range. But the wind isn't strong enough to carry sand with it. So the sand is left behind.

This process has been going on for thousands of years. By now there are tons and tons and miles and miles of wind-sculpted dunes there at the base of the high peaks.

Dunes you have seen might not have a mountain range to stop them. But at the start there was something there that made the wind drop its load of sand. Maybe it was just a clump of grass. Any little obstacle can cause the sand to collect. And once some sand is deposited, the sand itself becomes the obstruction. More and more sand piles up.

The winds that brought the sand there to begin with are still blowing, still bringing more sand. The winds lift the grains of sand on the windward side of the dune, pushing them up the slope. For a while, the sand piles on the dune's crest. But eventually the crest becomes overburdened and the sand starts sliding down the leeward side of the dune. Geologists call the flow of individual grains sand flow. When a large mass of sand slides down the slope, they call it slumping. Sand flow and slumping can move sand dunes forward many feet in a year.

There are places in the world where whole towns have been buried by creeping sand dunes. In the Sahara Desert of Africa, and at some locations in the Middle East, people live near oases in the deserts. They constantly shovel and scrape away the sand. They plant rows of date palms and water them with buckets of water drawn from the oasis well. They hope that the trees will stop the sand and keep it from burying their homes. But often it is the sand that wins the battle. Photographs taken by satellite cameras have revealed the outlines of ancient cities that now lie beneath hundreds of feet of sand.

Even in the United States, sand dunes have buried whole towns. Along the eastern shore of Lake Michigan, a string of sand dunes stretches for hundreds of miles. If you

The water and sand of Sleeping Bear Dunes National Lakeshore create a wonderful playground.

were to walk on these dunes, you might come upon an old chimney, or the peak of a roof, sticking up out of the sand.

Nowadays, people enjoy the dunes for what they are, beautiful sand hills, spotted here and there with clumps of beach grass, even some shrubs and trees. They have become a wonderful playground, a place for climbing, sliding, even sand-skiing.

Maybe someday you will travel around the United States to see, and play on, as many sand dunes as you can. Like those in Colorado or Michigan or Cape Cod, in Massachusetts. In North Carolina, you can climb up onto the dune where the Wright brothers made their first airplane flight.

In many places along the Pacific coastline, from California to Washington, there are dunes. And in New Mexico there are some of the most spectacular dunes in the world. As white as snow, they cover 275 square miles (712 square kilometers) of desert. Be sure you have sunglasses!

Across the border, in the Mexican state of Sonora, is a huge sand dune area called Gran Desierto. Spreading over 1,700 square miles (4,403 square kilometers), the dunes are the largest in North America, so large that they are referred to as a sand sea.

NATIONAL LAKESHORES

Along the shores of the Great Lakes, there are also beautiful beaches, some of them backed by high sand dunes. Four of these areas are preserved as National Lakeshores:

• Apostle Islands National Lakeshore, 20 islands off Wisconsin, in Lake Superior. 67,885 acres (27,473 hectares).

• Indiana Dunes National Lakeshore, in Indiana, on Lake Michigan's eastern shore. 12,870 acres (5,208 hectares).

• Sleeping Bear Dunes National Lakeshore, in Michigan, on Lake Michigan's eastern shore. 71,021 acres (28,742 hectares).

• Pictured Rocks National Lakeshore, in Michigan, on the shore of Lake Superior. 72,899 acres (29,502 hectares).

Plants of Desert Sands

If you have ever planted a garden, or even taken care of a houseplant, you probably know that most plants grow best in rich, dark soil. Yet many plants grow in sandy places, even those that may seldom get rain. Almost all deserts have plants growing in their gravelly or sandy soil.

Plants that grow on shifting sands must grow and produce their seeds very fast, before the sand shifts from under them or buries them too deeply. Many plants of the sand have very large seeds. That is because the larger the seed the more nourishment it can store. Small seeds, if they should become covered with deep sand, might not contain enough food to keep the young seedling alive until it can push its way up to the surface. Larger seeds store more food and can give a young plant more time in which to grow into the sunlight.

One of the most amazing things about the seeds of desert plants is how long they can lie in or on the ground, waiting for water to help them sprout and grow. Have you ever seen a desert area in the southwestern United States in bloom? Locations that may not have had more than a few drops of rain for many months, or even years, can suddenly come alive with millions of brightly colored blossoms after one heavy rain.

One of the most beautiful desert blossoms is that of the Engelmann prickly pear cactus.

Primroses in blossom after a summer rain.

Desert plants have learned other ways to adapt to their harsh environment. Cacti, for instance, do not have leaves, through which other kinds of plants lose a lot of water. They have thorns instead. Many cacti are thick and rubbery, with tissues that are able to store what little water they get.

If you have ever dug down into dry sand you may have been surprised when you came to sand that was wet. Sand is able to absorb whatever water falls on it. It doesn't let it run off, the way water runs off some other kinds of soil. Many desert plants have roots that grow deep into the sand to reach the water stored there. The deep roots of a single desert plant sometimes spread out over an area as big as a tennis court.

Desert plants have root systems that spread wide and deep in their search for water.

A Fishhook Mammillaria cactus growing in the Arizona desert.

Creatures of Desert Sands

Many of the world's creatures live in or on sand. A lot of them make their homes on beaches or in other sandy places near water. Some dwell on underwater sand, particularly those areas just off the coasts of continents, known as the continental shelves. These creatures are comfortably cool and have plenty of food and water.

But how do the creatures who live in hot, sandy deserts survive? They may be miles from water and they may not see rain for months at a time. The sands they run or crawl on are often so hot that, if you tried to walk on them barefoot, you would blister your toes. And walking on loose sand is not easy, with or without shoes on your feet. With each step, your foot digs into the sand and it takes extra effort to lift it up and out.

How, then, does a fringe-toed lizard, which spends a lot of its time running around on sand dunes, do it so easily? For one thing, the lizard is a lot lighter than you are, so it doesn't sink in as much. It also has extra large feet. Just as snowshoes can keep a person from sinking into soft snow, the broad scales on the lizard's hind feet keep it on top of the sand.

There is a much larger animal whose big feet help it walk on sand. That is the camel. Not only are its feet super-sized; they have thick pads on the bottom to protect them from the searing heat of desert sands.

Camels have other ways of coping with their sandy environment. They can keep plodding along, even through fierce sandstorms, because they have ways of keeping the blowing sand out of their eyes, ears, and nostrils. Their eyes are bordered by a double row of eyelashes, their ears are well fringed by hair, and they can close their nostrils whenever they want to.

The super-sized, well-padded feet of camels keep them from sinking into the sand.

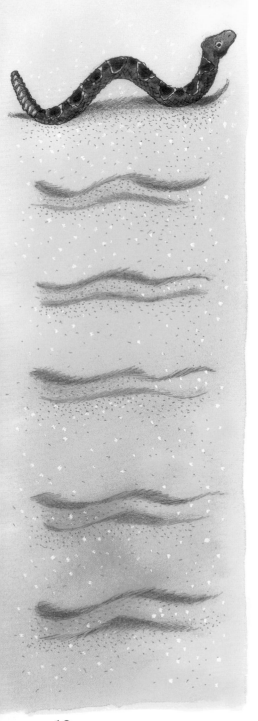

The sidewinder rattlesnake moves in such a way that only part of its body touches the hot sand at any one time.

To keep from burning their feet, some desert creatures move very fast. The roadrunner, the comical bird of the American Southwest, can run at a speed of 25 miles (40 kilometers) an hour.

Snakes can't move as fast as creatures with legs, so they have other ways to keep from getting scorched. The sidewinder rattlesnake tosses itself along sideways, in odd looping movements, so that only parts of its body touch the sand at any one time.

The kangaroo rat is one of the most amazing of all desert creatures, for in its entire life it never drinks water! It does gobble any green tidbits it can find, such as bites of succulent cactus, and from these it gets some water. But scientists have found that the kangaroo rat does not even need to eat blades of grass or cactus treats to get water. From the seeds it eats it gets hydrogen, and from the air it breathes it gets oxygen. You know what you can make by combining hydrogen and oxygen, don't you? Yes, water. Somehow the kangaroo rat's body uses these two elements to make all the water it needs.

Some desert creatures have found ways to put sand to work for them. The back-flip spider picks up grains of sand and weaves them right into its net. Then the spider flips onto its back and pulls the net over itself. Sooner or later the spider's next meal will come along. An insect, not recognizing the net for the trap it is, will get caught.

The kangaroo rat is well suited to desert living. In its entire life it never drinks water!

Many other creatures burrow into the sand to escape from both the heat and their predators. Some lizards actually swim in the sand. Transparent disks in their eyelids let them see where they are going even when their eyes are closed.

One creature, Grant's golden mole, never leaves its sandy underground burrows. It has neither eyes nor ears. How does it find its food? It is super sensitive to movement and can detect the small creatures it eats by vibrations they create in the sand.

Concrete from Sand

When you walk on a hard city sidewalk it may not seem much like walking on a soft sandy beach, but in both cases you are walking on sand. Sidewalks, buildings, bridges, dams, highways—concrete is used to build them all. And concrete is made with sand—mixed with gravel, water, and cement, of course.

Have you heard someone say "cement sidewalk" or "cement block?" That is not correct. The right word is concrete.

You may have seen cement and know that it is gray and powdery. It is made from ground, heated limestone and clay. When mixed with water, it makes the glue that turns sand and gravel into concrete.

Some concrete has more sand than gravel in it, and some has more gravel than sand. When there is more sand in the mixture, the finished concrete is smoother.

Builders aren't worried about running out of sand, but they do have a problem about digging huge sand and gravel pits near large cities. In shallow water off our coasts, there are millions of tons of sand. Many of the structures of the future will be built with sand from the sea.

When you mix water, cement, sand and gravel . . .

. . . like this, you get concrete. And you have to spread it out fast, before it hardens.

When concrete is needed for the building of highways, bridges and other huge structures, it is mixed by big machines.

Glass from Sand

Can you imagine a world without glass? Without windows, eyeglasses, mirrors, telescopes, camera lenses, light bulbs, jars, or bottles? No prism to hang in a window so you can watch a rainbow dancing on your wall? We have all these things because people discovered that glass can be made from sand.

Nobody knows for sure just how this discovery was made. We think it happened quite by accident about 6,000 years ago. Perhaps some people dug a pit on a sandy beach and in it built a fire. Hours later, they may have seen a strange, shiny lump among the ashes. If they had picked it up they might have been surprised at the way the light shone through it. Perhaps they had seen similar objects before, because glass had surely been made by lightning or meteorites striking sandy surfaces. But this time they may have realized that it was the intense heat of their roaring fire that had turned the sand into this mysterious, sparkling substance.

The earliest deliberately man-made glass we know about was manufactured about 3,000 B.C. in the lands along the eastern shore of the Mediterranean Sea. Some of the clay pots found there were covered by a glaze of glass. The earliest all-glass containers were probably made about 1,500 B.C. in Egypt and the area that is now Iraq.

Early humans may have wondered about the shiny substance they found after lightning struck a patch of sand.

When glass is made, the ingredients are first mixed and heated in giant furnaces to very high temperatures.

Ever since then, glassmakers have been mixing sand with other ingredients. The most common ones are two white, powdery substances—soda and lime. The mixture is then heated in a big furnace to very high temperatures. Different chemicals are added to make different kinds of glass. After reaching the melting point, the mixture is poured out of the furnace and shaped.

Once, many glass objects were made by glassblowers, people who took up a molten blob of glass on the end of a pipe and then blew through the other end of the pipe to shape the glass. Some glass art objects are still made this way. Mostly, however, molten glass is poured into molds and shaped into bottles, jars, and other objects, or it is rolled out to make flat glass, like that used for windows.

Beautiful windows are made of pieces of stained glass. This one was created about 1900 by the famous designer, Louis C. Tiffany.

While glass is molten it can be shaped into almost any form. This table was made in France in 1878.

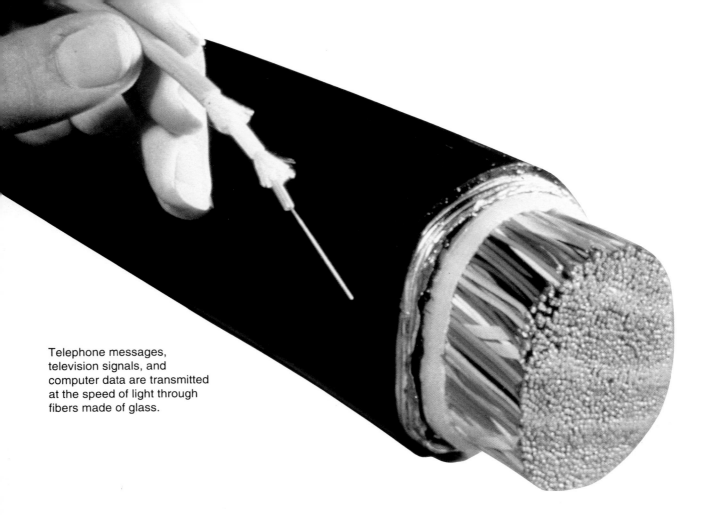

Telephone messages, television signals, and computer data are transmitted at the speed of light through fibers made of glass.

In the thousands of years that people have been making glass, many different kinds of glass have been created. One company alone makes over 100,000 kinds of glass!

One of the most amazing things about glass is that it can be turned into cloth. When liquid glass is forced through tiny holes it comes out in threads, each one being much thinner than a human hair. They can be woven into a kind of cloth called fiberglass. In your home, you may have curtains or draperies made of fiberglass.

Fiberglass can also be spun into glass wool, which looks something like cotton candy. It is used for insulation in the walls of ovens, refrigerators, buildings, airplanes, spacecraft, and even space suits. Without their glass-lined suits to protect them, our astronauts could not have survived on the moon's surface or during their walks in the fierce cold of outer space.

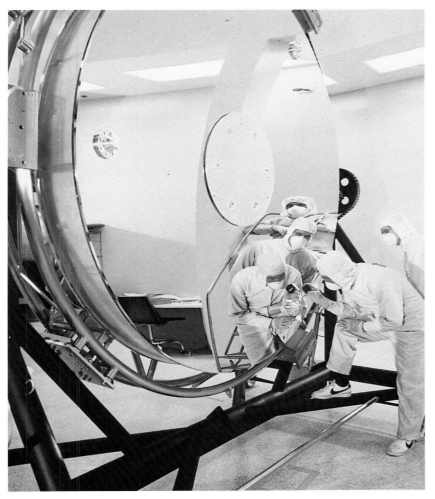

These technicians are inspecting a mirror for a NASA space telescope.

Suits lined with fiberglass keep astronauts warm in the cold of outer space.

SOME SPECIAL KINDS OF GLASS

By changing the ingredients and the way it is made, glass can be given different qualities. Here are some kinds of glass made for special purposes:

Laminated safety glass is a "sandwich" made of two layers of glass and one of plastic. If the outside glass breaks, the plastic layer holds the pieces together. Car windows and windshields are made of this kind of glass.

Safety glass is only one layer of glass, but it is given a special heat treatment that makes it very strong.

Bulletproof glass is made of several layers of glass and plastic.

Foam glass is very light because it is filled with tiny cells of gas. It is used as an insulating material.

Glass building blocks are two hollow half sections sealed together by heat.

Heat-resistant glass can stand great temperature changes without cracking.

Optical glass is a high quality glass used in eyeglasses, microscopes, telescopes, and camera lenses.

Glass optical fibers can bend light around corners. They are used to transmit telephone and television signals.

Laser glass contains fluorescent materials. It produces a narrow beam of light.

Photochromic glass turns darker when it is exposed to the sun. You may have worn sunglasses made of this special kind of glass.

Sand Paintings

In the American Southwest, Indian medicine men make sand paintings in their healing ceremonies. In the home of a person who is ill, the medicine man makes his painting on the bare floor. He uses sands of different colors. The Indians believe that by making paintings of their legendary gods and goddesses, they can bring them and their powers down from their homes in the sky. When a sand painting is finished, the medicine man sings a healing song.

Such sand paintings are always swept away when the ceremony is over. But American Indian artists, and other artists too, also make sand paintings that can be framed and kept. You, too, can make a sand painting. Here is what you need:

- A piece of wood, hardboard, or particle board
- A sheet of paper the same size as your board
- A pencil
- White glue
- A paper cup
- Sands of different colors
- Two fine-pointed paintbrushes
- Chalk or charcoal
- Acrylic sealer spray

You can buy colored sand at a craft store or a florist or garden shop. If you have some clean, white sand you can color it yourself. Put different colors of poster paint in jars, and add water and sand to each. Let the sand soak in the colored water for a while. Pour the water out and spread the sand out on newspapers to dry.

The next day put some white glue in the paper cup and dilute it with water so that it will be easier to spread. Coat your whole board with a thin layer of the glue.

Before the glue dries, shake sand over the board. Use whatever color you want to be the background color for your painting. Pat the sand down, fill in any bare spots, and let this layer dry overnight. The next day lift the board and shake off any loose sand.

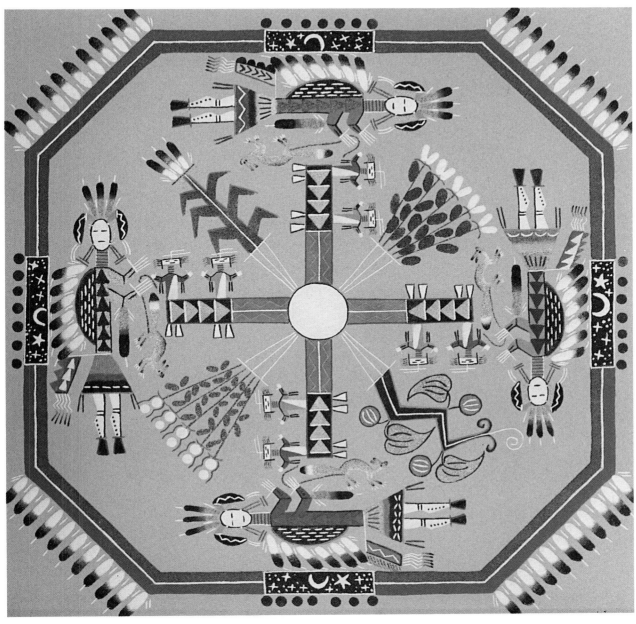

Indians of the American Southwest collect sands of many different colors to make their wonderful sand paintings.

In between these steps, you can be drawing your picture or design on the sheet of paper. With the tip of your pencil, punch holes along all the lines in your drawing. Then place your paper over the sandy board and, with chalk or charcoal, rub over the holes so that you can see the outlines of your drawing on your sandy board.

Dribble some glue on all the sections that are going to be one color—all the blue parts, for instance. With one of the paintbrushes, spread the glue evenly over them.

Now pick up some blue sand between your thumb and forefinger and sprinkle it over the areas where you spread the glue. When all the blue areas are filled in, pat the sand down and, again, let your painting dry for a while.

At any time you can lift your board to shake off any loose sand. You can also use your other paintbrush to brush sand away from places where you don't want it to be.

Continue to fill in the sections, one color at a time.

When your painting is finished, spray it with the acrylic sealer to make sure all the sand will stay in place.

Save whatever sand you have left for your next painting, or use it to make a different kind of sand art.

Maybe you have seen glass jars or vases filled with layers of colored sand that make pleasing designs or look like paintings of desert or mountain scenes. If you work

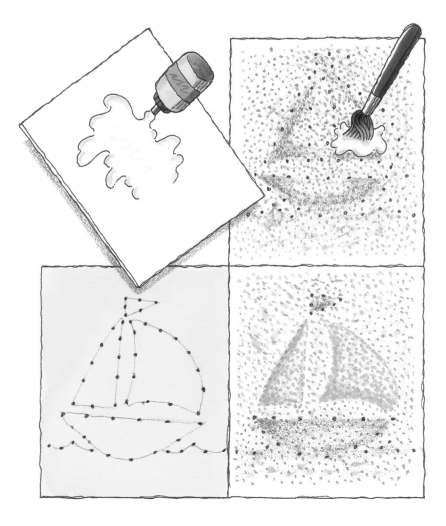

Top left: Cover your board with white glue. If the glue is too thick to spread easily, dilute it with water. Before the glue dries, sprinkle sand over the board.
Bottom left: Draw your picture on the sheet of paper and punch holes along the lines.
Top right: Spread glue on all sections that are going to be one color. Here the sails are being prepared for yellow sand.
Bottom right: While the glue is still wet, sprinkle on the second color.

slowly and patiently, this kind of art is not hard to do.

Choose a glass jar with a wide top and pour a layer of sand in the bottom. Carefully spoon sand of a different color on top of the first layer. Maybe you want this layer to look like a hill in a scene you are creating. Move the sand carefully with the spoon or a popsicle stick to pile it up against one side of the jar. Be careful not to dig down into the bottom layer. You can also use your paintbrush to push sand to where you want it.

You can add as many colors and layers as you want. You can fill your jar or bowl completely, or make just a few layers in the bottom half.

If you want to keep your artwork, you can seal the top of the sand by dripping a thin layer of wax from a burning candle onto it. Have an adult help you with this final step.

Sand Castles

There is one kind of sand art that you may have made yourself. When you were young, perhaps you filled a pail with wet sand, then turned the pail over and dumped the sand out. If the wet sand stuck together in a mound, you had the start of a little sand house or sand castle.

Now you can make a much bigger, and a much fancier, castle in the sand. Of course, the best place to do that is at the beach.

The next time you go to the lake or the seashore, take along some things you can use as tools. A ruler is very good for scraping flat walls. Popsicle sticks and plastic knives, forks, and spoons can be very useful too. You should also have a small shovel, especially if you want to start with a big pile of sand. And you definitely need a pail.

Start by making a pile of sand and wetting it down with pail after pail of water. Stamp around on your pile to pack it down. The more tightly sand is packed, the easier it is to carve and shape. Always work from the top down so that you don't scrape sand down onto parts that are finished.

Every fall sand sculptors gather at Harrison Hot Springs, in British Columbia, Canada, to see who can build the biggest and best sand castle.

Building sand castles is not just child's play any more. One team of sand artists constructed a 52½-foot (16-meter)-high "Lost City of Atlantis" on a California beach. Some people have made a business out of building castles, and other sculptures, in the sand. They get paid for building their creations at fairs, festivals, and conventions.

Every year there are competitions for sand sculptors. Judges cast votes to pick the best work of art. If you keep building sand castles when you go to the beach, maybe someday you will win a prize.

Some grown-ups get paid to create sculptures, like this one built at the New Mexico State Fair.

Index